The Management of End-Stage Chronic Heart Failure:

A Practical Guide

Both the author and the publisher make no representation, express or implied, that the drug dosages in this book are correct. Readers must therefore always check the product information and clinical procedures with the most up-to-date published product information and data sheets provided by the manufacturers and the most recent codes of conduct and safety regulations. The author and the publisher do not accept responsibility or legal liability for any errors in the text or for the misuse or misapplication of material in this work.

Whilst every effort has been made to ensure that the contents of this book are as complete, accurate and up-to-date as possible at the time of writing, the author or publisher are not able to give any guarantee or assurance that such is the case. Readers are urged to take appropriately qualified medical advice in all cases. The information in this book is intended to be useful to the general reader, but should not be used as a means of self-diagnosis or as the sole reference for the prescription of medication.

ISBN 978-1-4467-5948-6

The Management of End-Stage Chronic Heart Failure:

A Practical Guide

Dr Colin A J Farquharson *MBChB FRCP (Edin) FESC*

Consultant Cardiologist and Honorary Senior Lecturer in Cardiology

Northern Lincolnshire and Goole Hospitals NHS Foundation Trust, and Hull-York Medical School

United Kingdom

CONTENTS PAGE

DEDICATION

To Kay, my eternal soulmate, and Joel and Kristina

ABOUT THE AUTHOR

Dr Colin Farquharson is a Consultant Cardiologist at the Northern Lincolnshire and Goole Hospitals NHS Foundation Trust, as well as an Honorary Senior Lecturer in Cardiology at the Hull-York Medical School. He is also an Honorary Consultant Cardiologist at the Hull and East Yorkshire Hospitals NHS Trust.

He has a special interest in the management of Chronic Heart Failure. In 2001, he won the prestigious Menarini Academy International Clinical Cardiovascular Research Award in Rome, considered to have carried out the best cardiac research in the whole of Europe for his work on the treatment of Chronic Heart Failure.

INTRODUCTION AND GENERAL CONSIDERATIONS

In the United Kingdom (UK), between 1-2% of the adult population suffer from chronic heart failure (CHF), with approximately 6-10% of the elderly population (i.e. over 65 years of age) suffering from this disease. Although advances in the treatment of CHF have kept the age-adjusted incidence of the condition essentially stable, the overall prevalence has continued to rise with over 1 million cases found in the UK alone.

Despite the major advances in both the pharmacological management of CHF (using a combination of agents such as beta-blockers, angiotensin-converting-enzyme inhibitors, angiotensin receptor blockers and aldosterone antagonists) and the non-pharmacological interventions (such as cardiac resynchronisation therapy, ventricular assist devices, novel surgical techniques and cardiac transplantation), patients with advanced heart failure still remain a highly symptomatic cohort with a poor prognosis in general.

Thus, if all appropriate therapeutic options have been either used or discounted, then patients with advanced severe CHF represent an important and often difficult challenge with particular regard to end-of-life issues. Although cardiologists in general tend to be poor in changing the emphasis from aggressive intervention to palliative care and withdrawal of therapies previously thought essential, there is increasingly a change in mindset in considering that a death which is planned by the patient themselves along with their relatives with satisfactory symptomatic control and palliative care input is often a far preferable outcome than repeated hospitalisations for intravenous diuretics and inotropes taking up the majority of the patient's remaining lifespan.

This practical guide has therefore been prepared to provide a clinical focus for all health-care professionals who may have to provide supportive and palliative care to patients with heart failure with particular regard to troublesome symptom control towards the end of life.

The following is assumed:

- A full assessment of the patient's overall problems has been made, taking into account the comprehensive aspects of their physical, social, psychological, financial and spiritual issues, and appropriate referrals or actions have already been taken to address these. This should be carried out by the patient's usual attendant team: for example, by the heart failure nurse specialist if the patient has one, or by the general practitioner or hospital physician.
- The patient's heart failure management (including assessment and referral for cardiac resynchronisation therapy if appropriate) has been optimised and closely monitored.
- In the event of any significant symptomatic deterioration, any reversible factors have been detected and treated.
- There has been clear communication regarding the diagnosis and plan of management between not only the attendant team and patient / family and carers, but also between secondary and primary care.
- A clear plan for out-of-hours care is in place, particularly for patients with advanced disease who would be expected to continue to decompensate and deteriorate within a relatively short time-frame.
- Advice or referral is sought from the specialist palliative care team for persistent or complex problems, or for terminal care if the patient's stated preference is hospice-based care. A shared-care approach is advised, although the relative balance of inputs from cardiology and palliative care would alter as the patient deteriorates.

As the last days of a heart failure patient's life are often accompanied by considerable distress, discomfort and anxiety, it therefore follows that a cardinal principle of care would be to provide such a patient with satisfactory relief of pain, dyspnoea and other distressing symptoms and providing comforting measures in general. An equally important process is to negotiate an appropriate place to die (whether this be home, hospital or hospice), and to provide emotional and spiritual support for the patient and family, assisted by nurses and other appropriate health-care professionals.

REVIEW OF MEDICATION

When a CHF patient presents with an exacerbation of heart failure, or apparent "end-stage" heart failure, it is very important not to accept this diagnosis at face value. For every patient, it is important to exclude any reversible cause of symptomatic deterioration.

Common precipitants may include:

- **an episode of myocardial ischaemia**
- **onset of significant cardiac dysrhythmia**
- **non-compliance with heart failure medications**
- **concurrent respiratory tract infection.**

In these cases, treating the cause relieves the symptoms better than anything outlined below, although symptom-directed treatment may need to be given concurrently. In addition, it is always important to be certain that there is no potentially treatable or reversible cause for the heart failure itself. Always review the diagnosis and be certain of this fact: occasional patients with e.g. valvular heart disease and haemochromatosis can be rescued with specific interventions.

A thorough review of patients' medication is essential. For a full review of pharmacotherapy of heart failure, readers are advised to refer to current standard cardiology textbooks and up-to-date published heart failure guidelines. There are, however, a few basic principles specific to medical therapy for symptom management in end-stage HF. It is helpful for the attending clinicians to be clear which heart failure drugs are to be continued or can be withdrawn.

Drugs used fall into two categories:

a) **Those used as prophylactic therapy to improve the long-term heart failure prognosis**

b) **Those providing symptom relief and / or improvement in the patient's functional capacity**

The principles of continuation or withdrawal of medication for end-stage heart failure patients therefore consist in the following:

1. Drugs primarily prescribed to improve long-term prognosis can be withdrawn. The rationale for these changes needs to be discussed with patients and carers. Improving prognosis is subsidiary to improving symptoms in such patients.

2. Cardiovascular drugs primarily prescribed to improve symptoms or functional capacity (e.g. diuretics, digoxin or other digitalis analogues, and vasodilators) should be continued, but the dosages should be monitored, reviewed regularly and adjusted accordingly.

3. Always use the lowest doses of drugs necessary to produce the desired symptomatic benefits, discarding the established concept of targeting for "trial-proven doses".

4. The frequency and doses of drugs should be given to cover sufficient time periods. An inadequate regime resulting in frequent break-through symptoms can interrupt restful sleep and also be very distressing to the patient. For example, medication to prevent paroxysmal arrhythmias should cover the full 24 hour day, whereas medication to control exertional angina pectoris can be limited to cover the physically-active times of the day. Similarly, long-acting nitrates can be taken before bedtime, instead of the usual morning dosing regime, to alleviate nocturnal dyspnoeic symptoms or decubitus angina.

5. Weigh up the benefits versus the side-effects / adverse reactions – e.g. diuretics may impair renal function or precipitate gout; unwanted effects of Digoxin can also occur while drug levels are within the therapeutic ranges; excessive vasodilatation can lower blood pressure to cause presyncope / syncope / severe lethargy and fatigue.

6. The route of administration of any therapy requires careful consideration:

   a) The least invasive methods of delivery are preferable.

   b) The route of administration should also be balanced against efficacy. Oral administration may be inappropriate in patients with swallowing difficulties or in oedematous patients when intestinal absorption is often poor. Many patients with heart failure have gross oedema or poor dermal perfusion which may make subcutaneous or transdermal administration inappropriate.

   c) Personal and cultural sensitivities to some routes of administration (e.g. *per rectum*) should be respected.

   Common sense should always prevail, e.g. if the patient is exclusively mouth breathing, nasal cannulae can actually be placed into the mouth rather than the nostrils to deliver palliative oxygen therapy.

CONTINUING MEDICATION FOR IMPROVING HEART FAILURE SYMPTOMS

The key drugs to improve symptoms specifically arising from heart failure are diuretics, digoxin, vasodilators and inotropic agents. A brief guide is included here for the management of end-stage heart failure patients.

Diuretics

Despite the lack of large-scale randomised controlled trials, diuretics are considered the first line therapy for symptoms of congestion in patients with heart failure. Diuretics are often necessary because natural compensatory neurohormonal mechanisms in heart failure trigger fluid retention. Fluid retention and pulmonary congestion can be profoundly distressing and hence diuretics are usually necessary medication for end-stage heart failure therapy.

The aim of diuretic therapy in heart failure is to remove any excess fluid and then maintain fluid balance, ensuring fluid intake equals the overall output. Because either side of the equation can change independently, it is important to remember that diuretic dosages may need to be reduced if there is reduced fluid intake in order to prevent dehydration and progressive renal failure. Features of hypovolaemia or dehydration include dry mucosa, reduction of skin turgor, and postural / orthostatic hypotension. If fluid intake is significantly diminished fro any reason (e.g. nausea, vomiting, swallowing difficulties, sedation), then the diuretics may even need to be discontinued for periods of time.

In heart failure patients with peripheral oedema, it is not necessary to remove every trace of oedema by aggressive diuretic therapy. The margin between complete absence of peripheral oedema and substantial dehydration is narrow. Only symptomatic oedema or oedema-associated complications (e.g. painful ulcerations and pressure sores) require such aggressive diuretic therapy.

The particular choice of diuretics is up to individual clinicians, and contemporary management should be checked in standard up-to-date textbooks and review articles. It however is important to always look out for electrolyte disturbances which can cause arrhythmias with subsequent worsening of dyspnoea.

The general principles of diuretic dosing include:

- Avoiding nocturnal diuresis that will disturb sleep, unless the patient already has an indwelling urinary catheter.
- For patients with stress incontinence, the use of longer-acting loop diuretics (e.g. Torasemide) may be better tolerated and improve compliance.
- Diuretics are far more effective given intravenously than orally, often because impaired gastric absorption due to gut oedema. This is particularly true for Furosemide.
- Repeated split doses are more effective than a once-daily dosing regime.
- A continuous infusion of Furosemide (e.g. 10mg/hour) is more effective than repeated boluses to the same daily cumulative dose.
- If intravenous diuretic is not appropriate, then changing from oral Furosemide to Bumetanide often proves more effective due to improved gut absorption of the latter drug.
- Bed-rest will enhance diuresis, as will stopping aspirin therapy.

It is important to note that loop or thiazide diuretics can often cause profound hypokalaemia and hypomagnesaemia. Adding potassium-sparing diuretics (e.g. spironolactone or amiloride) are often better than potassium supplementation.

The use of combination diuretics requires special consideration:

- Combination diuretics can be very helpful where a patient requires very large doses of loop diuretic (more than 80mg Furosemide or its equivalent twice daily), or where a patient has become apparently resistant to loop diuretics.

- The combination of loop and thiazide diuretics (e.g. **Furosemide** *and* **Bendroflumethazide**) causes "sequential nephron blockade" and is often used to reinitiate significant diuresis, especially when IV administration is not considered an option. In many patients, regular loop and thiazide diuretic administration is needed to control fluid retention (see "self-management" below). The atypical thiazide diuretic **Metolazone** can be particularly effective in this situation when added to a loop diuretic, although one needs to be careful about avoiding dehydration and renal deterioration using this combination.
- Combination therapy may cause an abrupt and very profound diuresis, thus a patient's fluid balance should be carefully monitored. It is also particularly liable to upset electrolyte balance and is likely to cause hyponatraemia and deranged potassium values. Hyponatraemia may itself cause constitutional symptoms (such as cramps, confusion, seizures etc), and is usually a poor prognostic sign. However, low plasma sodium levels may be acceptable within the context of improved symptoms in end-stage disease.
- When starting combination therapy, it may take 48 hours or more before the maximal diuretic effects are seen.
- **Spironolactone** (or **eplerenone**, which is currently only licensed for use in heart failure following myocardial infarction) should usually be co-prescribed to reduce the risk of hypokalaemia.

Patients with late-stage heart failure may ultimately become diuretic resistant with severe symptomatic pulmonary and peripheral congestion. Measures that may help include the temporary use of inotropic drugs. A better approach can be ultra-filtration (a type of renal dialysis) which can allow large volumes of excess fluid to be removed rapidly. Once the oedema has resolved, provided there is still adequate renal function, the patient may become responsive to (and be able to be maintained on) regular oral diuretic therapy again. Ultra-filtration is however not widely available at present, and if when it is, good planning and discussion must take place before implementation.

Gout is another particular problem associated with using high-dose diuretic therapy. Non-steroidal anti-inflammatory drugs should usually be avoided to treat an acute attack as they cause fluid retention and worsen renal function. **Colchicine** is helpful in relieving symptoms, although nausea and diarrhoea usually limit medium and long-term use of this drug, and there should be a low threshold for starting treatment with **Allopurinol** (a xanthine oxidase inhibitor which also has beneficial effects in chronic heart failure) after an acute attack of gout has settled e.g. after 2 weeks symptom-free of gout.

Self-management of fluid balance

Many of the more motivated or able patients are keen to help manage their fluid balance, and this can be simply achieved, with a number of basic caveats. Bathroom scales and a supply of diuretics are all that is needed to adopt this approach:

- Determine the patient's target weight. For many patients with severe heart failure, this will be a weight that includes some excess fluid (i.e. leg oedema) below which significant renal impairment occurs. The patient should then weigh themselves daily. If the patient is fluid overloaded, limit fluid intake to approximately 1500ml / day.
- If weight increases by more than 2kg from baseline over 3 days, the patient should take an extra daily diuretic (generally by increasing background loop diuretic) until back to target weight.
- Conversely, if weight falls 2kg or more below target, the patient should omit half of that day's diuretic until back to target
- If weight changes by more than 5kg from target, the patient should seek medical advice and be closely monitored.

Digoxin

In selected patients with severe systolic dysfunction, digoxin may reduce symptoms and improve cardiac function. It is particularly effective in patients with atrial fibrillation as it helps control ventricular rate, and is frequently helpful in patients in sinus rhythm. Toxic effects (such as nausea) are more likely to be seen at lower doses when renal function is impaired. **Avoid hypokalaemia** in patients taking digoxin as it can precipitate toxicity even at low doses.

Nitro-vasodilators

Nitrates have little role in treating congestion as nitrate tolerance (tachyphylaxis) develops very rapidly and they are poorly absorbed after oral administration. They can be helpful in patients with acute symptoms, such as in paroxysmal nocturnal dyspnoea (PND) when administered sub-lingually. In patients with acute pulmonary oedema, intravenous infusion of nitrate is generally advised. Nitrates may help in patients with angina, but generally better anti-anginal drugs are available. The rapid development of tolerance means that a nitrate-free period of 8 hours out of every 24 hours is necessary. Nitrates are therefore probably best used to address specific problems, such as nocturnal angina or frequent PND, when a long acting nitrate given at night may be helpful to try and give the patient a good night's sleep.

Morphine

Morphine can be effective for pain, cough and dyspnoea in patients with heart failure. It should be prescribed regularly *and* as required, with the breakthrough dose of e.g. Oramorph being one sixth of the total 24 hour dose of Morphine. A usual starting dose of Oramorph would be 2.5mg - 5 mg every 4 hours and as required, using 2.5 mg initially if the patient is frail.

Common problems on initiating morphine are nausea and drowsiness, which both tend to resolve within a few days. The nausea may need short term treatment with e.g. **Haloperidol** 1.5mg – 3mg nocte. **All patients on morphine need laxatives** (see Constipation section *vide infra*), which should be co-prescribed. If the patient is already on morphine for pain, a dose increase of 30-50% may help with dyspnoea.

Active metabolites of morphine and other opioid drugs accumulate in renal impairment. In these cases, the frequency of administration should be reduced, i.e. prescribe Oramorph either thrice daily or even twice daily (and as required). Avoid slow release opioid preparations in renal impairment if at all possible.

If problems occur with morphine, alternative preparations are available, although it is often sensible to discuss these specifically with local palliative care specialists.

Positive inotropic drugs

Most positive inotropic agents have to be administered intravenously, usually making them unsuitable for palliative-care stage heart failure. However, symptoms may be severe and intractable despite all the other best available medical therapy. In such situations, a few hours of inotropic support may provide much needed symptomatic benefit to the patient. Inotropic drugs all appear to worsen prognosis, and therefore should only be used in this setting to treat intractable symptoms. Only Dobutamine can be given intravenously via a peripheral vein (in dilute form) versus via central venous access, which all other agents require because of their highly-irritant nature and vasoconstrictive actions peripherally. In end-stage disease, there should be careful patient selection and a clear intended benefit, with plans made to review this regularly.

GENERAL PRINCIPLES FOR THE WITHDRAWAL OF CARDIOVASCULAR DRUGS IN PATIENTS WITH END-STAGE CHRONIC HEART FAILURE

Polypharmacy is commonly prevalent and burdensome in palliative care. The burden of dual administration versus the benefits should be considered for each individual. As each patient will be different, this information is designed to be a guide rather than an absolute rule about the order in which to consider reducing therapy.

Cholesterol lowering drugs (e.g. statins) can usually be the first to be discontinued because they have no symptom-relieving properties (and have no proven prognostic benefit in heart failure).

Anti-arrhythmic drugs, including beta-blockers, can also be considered for discontinuation at an early stage. Most anti-arrhythmic drugs lower blood pressure and can contribute to symptoms of fatigue. If symptomatic tachycardia occurs, or if a drug such as a beta-blocker is also helping angina symptoms, it may be best to continue this therapy specifically. In some patients, Amiodarone can control otherwise very symptomatic arrhythmia, such as atrial fibrillation with a rapid ventricular response or episodes of ventricular tachycardia.

Anti-anginal agents can be discontinued if the patient has no troublesome angina. Antihypertensive drugs are usually also inappropriate in end-stage heart failure.

If possible, continue with ACE-inhibitors / angiotensin receptor blockers as they do provide some symptomatic relief in end-stage heart failure. However, stop if symptomatic hypotension, cough, worsening renal failure or the quantity of medications taken is troublesome to the individual patient.

Diuretics should be continued as long as possible, including loop diuretics (available in liquid form), thiazides and Spironolactone, in order to reduce the recurrence of symptomatic fluid congestion.

Anti-platelet and anti-coagulant therapy should however be considered on an individual basis, based on the risks and benefits in each individual patient.

- Antiplatelet agents are typically given to lower the risk of ischaemic events (ACS / MI). They have never been shown to improve prognosis in heart failure and aspirin in particular is associated with an increase in the risk of hospital admission in patients with heart failure. Aspirin also antagonises the effects of many diuretics and contributes to diuretic resistance. It may therefore be stopped if angina is not prevalent in the patient with end-stage heart failure.
- Warfarin is mainly used to reduce the long-term stroke risk, particularly in patients with atrial fibrillation. The decision to stop warfarin therefore requires more discussion. Some patients, not unreasonably, prefer not having repeated blood tests to monitor INR control. Subcutaneous low-molecular weight Heparin (low or full dose) can be an alternative if anticoagulation in the end-stage setting is still thought to be clinically appropriate.
- If a patient with HF is bed-bound, then their risk of venous thrombosis and pulmonary embolus is extremely high. Such a patient may be offered prophylaxis with low-dose subcutaneous low-molecular weight heparin.

ANXIETY AND DEPRESSION

It is thought that at least one-third of all heart failure patients suffer from some form of depressive illness. The psychological distress that accompanies depression impacts adversely on many symptoms, the patient's subsequent understanding of their disease, compliance with potentially helpful medication and overall survival.

A full assessment of contributory factors and involvement of the multi-professional team as needed is often required to address this sometimes difficult issue to resolve.

Non pharmacological approaches:

- Consider referral for cognitive behavioural therapy
- Consider low-impact exercise programmes
  - Encourage social stimulation of available and suitable activities – this may include day hospice attendance and heart failure support group attendance
  - Consider breathing exercises / training, with particular regard to reducing anxiety / panic breathing

Pharmacological approaches:

Depression

- Avoid tricyclic antidepressants and venlafaxine in general, in view of potential cardiotoxic side-effects.
- Selective serotonin reuptake inhibitors, such as **Sertraline** 50mg daily, **Citalopram** 20 - 40mg daily and **Mirtazepine** 15 - 30mg nocte appear to be relatively "heart safe". Sertraline is currently recommended as the first line treatment in active ischaemic heart disease.

- **Citalopram** has fewer drug interactions, especially with warfarin.
- **Mirtazepine** may be useful if there is also persistent nausea or poor appetite in cachetic patients.
- **Trazadone** (starting at 50mg nocte, and titrated as tolerated to a maximum of 200mg nocte) may be useful if insomnia is also a problem.

Anxiety

- If panic is a major feature of breathlessness, then **Lorazepam** 0.5 - 1mg may be helpful and can be absorbed sublingually. Care should be taken, however, because of increased risk of falls and memory loss with benzodiazepine use, particularly in elderly patients with heart failure.
- For persistent anxiety, **Fluoxetine** 20 - 60mg daily (20 - 40mg in the elderly) or **Citalopram** 10 - 20mg at night may be more appropriate.
- **Diazepam** 2 – 5 mg orally at night can also be helpful for nocturnal anxiety causing insomnia. The effects of diazepam are longer acting than Lorazepam – may be effective for more persistent anxiety symptoms.

BREATHLESSNESS

It is certain that breathlessness is the major and most distressing symptom of advanced heart failure. Effective management of this symptom is thus dependent on accurate understanding of the nature of dyspnoea and the diagnosis of the mechanisms responsible. Differentiating pulmonary from cardiac causes of dyspnoea is usually the first step, as overlaps between the two types of cause are not uncommon.

Acute deterioration almost always has an underlying (and therefore potentially treatable) trigger. Substrates that should be considered include:

- **Chest infection/ pneumonia**
- **New-onset atrial fibrillation (with / without a rapid ventricular response)**
- **Pulmonary Embolus**
- **Failure to comply with medication**
- **New-onset / worsening myocardial ischaemia**
- **High salt load**

Once it is clear that the patient's breathlessness cannot be improved by further changes in heart failure therapy and that there are no other correctable causes, one or more of the following palliative measures can be considered:

Non-pharmacological management

- Optimise rehabilitation potential with an appropriate exercise programme to address cardio-respiratory deconditioning and maintain muscle bulk
- Breathlessness management, including breathing retraining, pacing / prioritising and panic management for anxiety attacks
- Occupational therapy to enable maximal mobility and functioning
- Psychological support – this may be provided through the nurse specialist or may require other referral including day hospice if available
- Relaxation therapy

- Complementary therapies
- Hand-held fan
- If night time breathlessness is a problem due to the patient slipping down the bed then using a back raiser, raising the foot of the bed mattress with wedges or even a profiling bed in the home may be helpful

If the patient reports daytime sleepiness and waking at night with gasping (or sleep partner reports long recurrent apnoeic spells), consider the presence of sleep disordered breathing. A referral for nocturnal continuous positive pressure airways pressure may be appropriate although not all patients can tolerate this. Central sleep apnoea with sleep disordered breathing presents in 50% of end-stage heart failure patients.

Pharmacological management

- **Oxygen**: Stable heart failure patients do not desaturate unless there is sleep disordered breathing, or a precipitating cause. The latter should be sought and treated. Breathlessness will not be relieved by oxygen in the patient with normal arterial oxygen and also may not in those with hypoxia (although other problems such as hypoxia-associated confusion may be helped). There is evidence that oxygen is harmful in the context of heart failure and normal saturation, and it should ONLY be used in a patient with demonstrated hypoxia. Humidified oxygen starting at 24% should be used with caution and as an empirical trial with a clear review to evaluate benefit. If sleep disordered breathing is present, then specialist respiratory advice should be sought. Do not exceed this concentration if there is co-existent COPD unless patient is known not to retain carbon dioxide. Nasal cannulae are often more acceptable to patients than facemasks which can be claustrophobic.
- **Short-acting Nitrates:** GTN Spray 1 - 2 puffs as required. Nitrates can be helpful in acute episodes of breathlessness because of their rapid onset, but are contraindicated in severe aortic stenosis. They may cause dizziness, especially if pre-existing orthostatic hypotension.

- **Longer-acting nitrates.** Usually **Isosorbide Mononitrate MR** 30 – 120 mg once daily, or alternatively transdermal GTN patches (10 – 20 mg) or **Buccal Suscard** 1 - 3 mgs at night for patients unable to swallow tablets
- *It is important to provide at least 8 hours of nitrate-free period in each 24 hours to avoid tachyphylaxis*
- **Nebulised saline 0.9% +/- bronchodilators:** e.g. **salbutamol** 2.5 mg or **terbutaline** 2.5 mg up to four times daily if significant concurrent COPD. If co-existing angina, GTN spray should be available if using nebulised sympathomimetic bronchodilators as angina may be precipitated by the resulting increase in heart rate.
- **Low-dose morphine sulphate oral solution**: The initial dose is suggested at 2.5mg four times daily and titrated accordingly. The possible mechanisms of action include reduction in central perception of breathlessness (similar to reduced central perception of pain), reduction in anxiety, reduction in sensitivity to hypercapnoea, reduced oxygen consumption and improvement in cardiovascular function. It is likely that the influence of different mechanisms varies in different people. A dose of opioid at night may ease sleep disordered breathing and improve resultant nocturnal breathlessness and sleep disturbance.

  Morphine is excreted renally, so if renal impairment / failure is present, use lowest dose initially and reduce frequency to twice or thrice daily depending on response. Sustained release preparations should be used with caution in the presence of renal impairment. For patients who are sensitive to morphine, alternative opioids may be suitable. When using regular opioids, prophylactic laxatives are strongly recommended and access to a suitable anti-emetic, as some patients may feel nauseated initially.

- **Sublingual Lorazepam:** e.g. 0.5 – 1mg as required (max 2mg / 24 hours). The rapid onset of action makes this useful in panic attacks, but should be limited to this situation because of concerns regarding falls and cognitive impairment.
- **Oral Diazepam**: e.g. 2mg – 5 mg bd is a longer acting agent. This effect may be useful but it can accumulate in hepatic impairment.

CONSTIPATION

Constipation may be caused by:

- **Reduced fluid and food intake**
- **Diuretics – causing relative dehydration**
- **Immobility**
- **Opioid medication**
- **Use of calcium antagonists, particularly verapamil**

It is therefore very important indeed to prescribe prophylactic laxatives in patients with end-stage heart failure, particularly when starting opioid preparations. The type of laxative used however is important to consider, and the following principles are relevant:

Unless there is a previous diagnosis such as irritable bowel disease, avoid bulk forming agents (such as ispaghula husk) if at all possible, because fluid restriction makes it difficult to drink the requisite daily recommended amount required to use these drugs.

Movicol can be used without any significant problems in end-stage heart failure. **Senna** may be required in addition as a stimulant laxative (starting dose is usually 2 tablets nocte), or **Docusate** as an osmotic agent (starting dose 100mg twice daily).

Co-danthramer is a stool softener and bowel stimulant that may be useful if the patient is still constipated despite using a polymacrogol with / without senna, and is only licensed in terminal illness. A typical starting dose would be two capsules twice daily or 10mls of liquid twice daily. If the strong suspension is used (i.e. 75/1000), this should be started at 5ml once or twice daily.

COUGH

Productive Cough

In patients with troublesome productive cough, it is important to consider the usual causes of cough, such as chest infection or worsening pulmonary oedema.

Non-Productive Cough

Cough due to angiotensin-converting-enzyme inhibitor (ACEI) drugs can begin some time after commencement, therefore it could be appropriate to attempt a trial of withdrawal of ACE inhibitor for at least 1 week, even if patient has been taking it for some time. Consideration could also be given to replacing the ACE inhibitor with an ARB (angiotensin receptor blocker) drug.

If cough continues despite the above measures, consider the following:

- If difficulty in expectorating sputum:

  Nebulised Sodium Chloride 0.9% as required. Usual dose 2.5ml – 5ml.

  NOTE: this dose accounts for 23mg – 45mg of sodium chloride. However, the amount of sodium chloride absorbed from the pulmonary system is so small that it will have no clinical consequence

- Cough suppressants for dry cough: **Codeine linctus** 5mls – 10 mls as required, to a maximum of four times daily.
- Low-dose **oral morphine** (e.g. **Oramorph** - starting dose 2.5 mg every 4 hours as tolerated) as required. This may also help with breathlessness and pain

  NB: *Always consider use of prophylactic laxatives when commencing strong opioids*

Consider other causes:

- Gastric reflux if nocturnal – a therapeutic trial of a protein pump inhibitor may alleviate this (e.g. **Omeprazole** 20 – 40mg at night)
- Late-onset cough-variant asthma – may be alleviated by inhaled bronchodilator / corticosteroid therapy
- Nocturnal pulmonary oedema – transdermal GTN patches or oral long-acting nitrovasodilator therapy may reduce this.

DRY AND / OR ENCRUSTED MOUTH

Ensuring satisfactory mouth care to promote good oral hygiene can make a substantial difference to quality of life and impact on other symptoms such as nausea, anorexia and weight loss.

Common underlying causes for mouth problems include dry oxygen therapy (ensure this is humidified if at all possible), medication, and underlying oral candidiasis (thrush).

Assessment should be carried out for other underlying causes e.g. excessive diuretic dosage, opioid use, anticholinergic or other saliva-reducing drug therapy.

If present and troublesome, consider trial of:

- Ice cubes / crushed ice / ice lollies
- Sipping Pineapple juice / sucking pineapple chunks (contains papain)
- Chewing gum (Sugar-free gum may reduce risk of oral candidiasis)
- Orange or satsuma segments
- Oral Balance gel
- Glandosane artificial saliva

Grapefruit juice can be used, but *with caution* as this can interact with drugs such as statins and some calcium-channel blockers (but not amlodipine).

EXCESSIVE FATIGUE AND LETHARGY

These are some of the most common and difficult symptoms to treat in end-stage heart failure. It is often unfortunately irreversible, but it is very important to look for potentially reversible causes which might include:

- Low cardiac output or hypotension: benefit may be possibly gained by the addition of digoxin, or reduction of dosages of Beta-Blocker, ACE Inhibitor / ARB and / or Diuretics. Sometimes it is necessary to reduce medication, which is of proven clinical benefit, because the adverse effects of hypotension and fatigue are increasing unacceptable to the individual patient.
- Hypovolaemia secondary to excessive diuretic dosages: check skin turgor, blood biochemistry including urea and electrolytes, as well as assessing postural hypotension – then adjust diuretic dose and frequency accordingly
- Anaemia: consider investigation / discuss empirical correction e.g. intermittent Iron infusion, oral iron therapy, blood transfusion etc.
- Hypokalaemia / hyperkalaemia: check urea and electrolytes and correct any electrolyte imbalance appropriately
- Hypothyroidism / diabetes: check thyroid function tests and blood sugar levels
- Hyponatraemia: this may respond to an adjustment of diuretic dose, but it may simply reflect a feature of end-stage disease and signify a poor prognosis
- Cardio-respiratory deconditioning: refer for low-impact exercise rehabilitation programme if available

Other things to consider may be patient lifestyle adaptation, formal Occupational Therapy assessment with a view to providing aids to assist activities of daily living, and Clinical Psychologist review to provide assistance with coping strategies.

It may also be worthwhile for further Cardiology review to assess for progressive cardiac dyssynchrony, as benefit may be gained from the implantation of cardiac resynchronisation devices such as biventricular pacemakers.

INSOMNIA

This is a common but often missed symptom in patients with end-stage heart failure. Look for contributory factors, such as:

- **Anxiety and depression**
- **Breathlessness / Central sleep apnoea**
- **Difficulty with positioning / Reduced mobility in general**
- **Pain**

Paroxysmal nocturnal dyspnoea (PND) is the symptom of waking in the night with shortness of breath. Patients who have previously experienced episodes of PND may have anxieties about its recurrence. This can have an impact on both the patient's and the carer's ability to experience restful sleep. These anxieties should be addressed on an individual patient basis and practical advice about management of PND should be offered.

- When there are signs of fluid overload, an afternoon or early evening dose of diuretic, and an evening dose of a long-acting nitrate, may be beneficial (see section on breathlessness *supra vide*). Use of GTN spray can give immediate, short-term relief (but avoid use in severe aortic stenosis).
- Relaxation and anxiety management
- If central sleep apnoea is problematic, a small dose of an opioid, such as **Oramorph** solution 2.5mg - 5mg at night, or night oxygen therapy may help
- Night sedation may be required: **Temazepam** 10mg - 20mg at night, or **Lorazepam** 0.5mg - 1mg at night, or **Zopiclone** 3.75mg - 7.5mg at night or **Trazadone** 50mg at night
- If associated with nausea, **Haloperidol** 1.5mg - 3mg at night, or **Levomepromazine** 6mg - 12.5mg at night may be specifically more helpful.

Disturbance due to slipping down the bed at night may be helped by a profiling bed. Other measures such as raising the mattress at the knees, using a back raiser or propped pillows may also be beneficial.

NAUSEA AND VOMITING

Patients with heart failure may have multiple causes of nausea and vomiting, which include gut oedema, hepatic engorgement and renal failure.

If vomiting is significant, or if there is gut oedema, the oral route may be temporarily ineffective and the subcutaneous (SC) route is therefore advised – either by stat injections or by continuous infusion via a syringe driver. When vomiting is subsequently controlled, the oral route may be reconsidered.

- Consider side-effects of medication as a cause of nausea and vomiting
- If nausea becomes constant, or there is renal impairment: try **Haloperidol** 1.5 - 3mg orally or sc (especially at night)
- If related to meals, constipation, early satiety, vomiting of undigested food, or hepatomegaly due to congestion: **Metoclopramide** 10 - 20mg thrice daily orally or **Domperidone** 10 - 20mg thrice daily orally or subcutaneously
- If nausea and vomiting persists despite the above: consider e.g. low-dose **Levomepromazine** 6mg nocte (the usual tablet preparation is 25mg, but pharmaceutical companies do provide a scored 6mg on a named doctor and patient basis if 3 - 6mg is required)
- Avoid cyclizine if at all possible, as this may worsen heart failure

NB: It is recognised that some of the drugs recommended above either have a potential to prolong the QT interval or act as an anti-cholinergic – however in the doses recommended, these adverse effects are thought to be minimal and to be appropriate in the symptom management of these patients with end-stage disease.

PAIN CONTROL

Patients require a full assessment of pain, including the site of discomfort and possible reversible causes.

Remember to consider other causes and pathologies e.g. angina pectoris and musculo-skeletal pain in addition to heart failure.

- NSAIDS should be avoided as they may exacerbate heart failure

The World Health Organisation analgesic ladder, although developed for cancer patients, can be applied equally to patients with heart failure and is appropriate particularly for the commonly experienced musculo-skeletal pain.

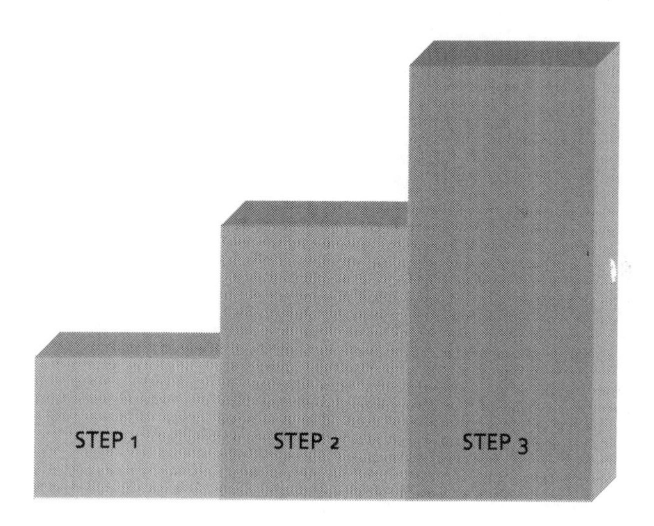

Step 1: **Simple analgesia** e.g. **Paracetamol** 1G four times daily regularly

Step 2: Add a **weak opioid** to Step 1 e.g **Codeine** 30 - 60mgs mg plus **Paracetamol** 1G four times daily, or **Paracetamol** 1G with **Tramadol** 50mg four times daily

***Step 3*: Strong opioid**. Stop the weak opioid commenced in Step 2, and replace with e.g. oral **Morphine Sulphate** solution 2.5mg four times daily and titrate as necessary. Reduce dose frequency in renal impairment. If renal function is markedly impaired, alternative opioids may need to be considered.

Dose titration can be carried out by increasing the regular oral morphine dose in steps of approximately 30% (or according to breakthrough doses required) until pain is controlled or side effects develop.

This can then be converted to modified release morphine when stable. The 24 hour dose of immediate release morphine is divided by 2, and then this dose can be prescribed as modified release oral morphine 12 hourly. A breakthrough analgesic should be prescribed as one-sixth of the total 24 hour morphine dose.

- Ensure that the Laxative dose is also increased as needed

Subcutaneous (SC) opioid analgesia

- Usually given in a syringe driver / pump over 24 hours
- Calculate the 24 hour total dose of oral morphine
- Convert this to either SC morphine or diamorphine using following ratio
    - Oral morphine 2mg = SC morphine 1mg
    - Oral morphine 3mg = SC diamorphine 1mg
- Prescribe one-sixth of the total 24 hour SC opioid dose as required SC for breakthrough pain

Other considerations

- With pain possibly related to recurrent myocardial ischaemia, anti-anginal medication should be optimised
- Hepatic pain may be due to hepatic engorgement with capsule stretch, and this pain may respond to nitrate therapy to reduce portal hypertension

- Gout is common in heart failure patients. For an acute attack, **Colchicine** is the treatment of choice. **Allopurinol** should be used to prevent further attacks – use a lower dose in renal failure. Although steroids are also relatively contraindicated in heart failure because of fluid retention, sometimes a short course of **Prednisolone** (e.g. 10mg daily) or **Dexamethasone** (e.g. 4 - 12 mg daily) may help to control an acute attack of gout. Careful monitoring of daily weight should alert to significant fluid retention. For mono-articular gout, intra-articular injection of steroids may be helpful.
- TENS machines can be helpful in some circumstances, but should be used with caution in the presence of automated implantable cardioverter defibrillators and should be discussed with a cardiologist first.

PERIPHERAL OEDEMA

Firstly assess whether this is due to any other cause that may be potentially reversible or treated with a different treatment strategy e.g.

- **Hypoalbuminaemia**
- **Liver failure**
- **Venous insufficiency**
- **Varicose veins**
- **Calcium-channel blocker therapy e.g. Verapamil**

Peripheral oedema in heart failure may also be secondary to right heart congestion. Complications include leg ulceration, bedsores, stasis eczema, and cellulitis. Peripheral oedema ranges from very mild, dependent ankle oedema occurring only in the evenings, to very severe associated with ascites, scrotal congestion and sub-conjunctival oedema. The objective of treatment is to relieve the associated symptoms and prevent complications, such as:

- Optimise diuretic therapy
- Cellulitis and leg ulcers are common and can be managed by the district nursing team who can seek further advice from the tissue viability nurse specialist if required. However, persistent and painful cases, especially those complicated by infected ulceration, may need prolonged antibiotic treatment.
- Compression leg bandaging or a scrotal support may be helpful, particularly in the presence of lymphorrhoea
- Good skin care is mandatory - *aqueous cream* as a barrier may be helpful
- Pressure relieving equipment may be needed e.g. a pressure relieving mattress, and / or profiling bed

ANOREXIA, WEIGHT LOSS AND CACHEXIA IN END-STAGE HEART FAILURE

Patient with HF may have a poor appetite and lose significant amounts of weight. Poor appetite may be exacerbated by breathlessness, fatigue, oedema, drug side-effects, renal dysfunction and depression.

In early disease, assessment of reversible issues, nutritional assessment and dietary advice may help the patient regain weight and improve quality of life. However, at end-stage, the neurohormonal syndrome of anorexia-cachexia may render dietary supplementation futile and burdensome for the patient.

There is currently no specific evidence-based guidance, and the following are based on best empirical practice for end-stage disease:

- Look for easily reversible factors and address them
- Dietary messages can be confusing as patients may be following cardio-protective dietary advice, using low fat foods which are too low in energy for their current needs. Patients who increase their nutritional intake and prevent further weight loss, or increase their "dry" weight may have an improved sense of well being and improved body image.
- There may be family expectations relating to food intake that can make mealtimes stressful. In general, give permission for the patient to eat small frequent meals of whatever they want, whenever they want. This may help with nausea and bloating.
- Relaxation of a low salt diet (if unpalatable) is often helpful.
- Assess the patient's need for assistance with shopping or cooking.
- Some patients may benefit from oral nutritional supplements, but if the patient finds them a burden they should not be insisted upon. Referral to a Dietitian for advice regarding the use of prescribed dietary supplements may be helpful.
- Small amounts of alcohol / steroid can be used as an appetite stimulant.

WITHDRAWAL OF AUTOMATIC IMPLANTABLE CARDIAC DEFIBRILLATOR / DEVICE THERAPY

Automatic Implantable Cardioverter Defibrillator (AICD) therapies are increasingly commonplace in the management of heart failure, especially those combined with cardiac resynchronisation therapy (CRT).

AICD devices are programmed to sense malignant ventricular dysrrhythmia and deliver a shock with the aim of restoring sinus rhythm and prevent sudden cardiac death (SCD). They are associated with increased survival in patients with NYHA class I or II symptoms.

However, as end-stage disease approaches, the myocardium will become less responsive to electrical cardioversion. In this setting, the concern is that the patient's death will be accompanied by the AICD discharging unsuccessfully, or more distressingly, successful electrical cardioversion to sinus rhythm only being short-lived, leading to repeated clear conscious painful discharges in the time prior to the patients' eventual death. This would be distressing to patient, carers and staff.

Patients may have very different attitudes to their AICD from fear of its discharge to fear of it being "taken away", not appreciating that a time will come when it will not be able to successfully restore normal rhythm. They may also have erroneous beliefs, equating "reprogramming to pacemaker mode" to "immediate death" and the patient may need to be given reassurance about this issue.

The decision to deactivate an implantable defibrillator can be difficult for patients and their relatives, and this should be addressed on an individual basis. Patients often report a perceived dependence on the device.

(The final cause of death in severe heart failure patients is most commonly due to progressive heart failure or from tachyarrhythmia).

General principles

- Eventual withdrawal of AICD care should be discussed prior to initial implant in all AICD recipients. Where this has not happened, this subject should be discussed as early as possible.
- It is appropriate to deactivate AICD devices in patients with end-stage heart failure and constitutes withdrawal of treatment. Not all patients will have a tachyarrhythmia in the terminal stage, but it is humane to try and avoid multiple shocks in those who do.
- Deactivation of the device in the community setting is problematic. It should be considered and discussed at the same time as "do not resuscitate" decisions are made and ideally be discussed if a patient is being discharged from hospital to palliative community care. Some functions can be turned off after discussion with the patient and family.
- When anti-arrhythmic medical therapy is being withdrawn, patients should be aware that device activation is then more likely and they may consider switching off the defibrillator function in anticipation of this.
- AICD patients should be encouraged to express their concerns especially in relation to their mode of death and shocks.

There should be liaison with the relevant defibrillator clinic and specialist nursing staff, along with the supervising electrophysiologist / device physician when deactivation is being considered. Local systems should be in place for deactivation in the patient who is too unwell to attend the clinic where it was inserted. Local device technicians will need to know the type of device used: the patient may have a card with the make on it, or the clinic which placed the device will have a record.

- Magnets will *temporarily* deactivate AICD devices whilst being held directly over the AICD and should only be used in the emergency situation. The magnets required are of a significant size and are specialized and expensive. Most coronary care units or cardiac physiology departments will be able to advise.

- When a patient dies with an active AICD, local systems should be in place for cardiology staff to deactivate the device before removal by mortuary or undertaker staff. Cremation cannot take place with an AICD in situ.

Where the patient is entering end-stage heart failure and AICD deactivation is being contemplated, it should be explained that:

- Deactivation of their AICD device does not mean that they will die imminently.
- The AICD may have been of value in prolonging their life in the past, but it may no longer be in their best interest for the patient to receive painful and often traumatic shocks.
- Pacing functions including Cardiac Resynchronisation can be left active (with the defibrillator mode inactive).

TERMINAL END-STAGE HEART FAILURE: THE LAST FEW DAYS

A significant percentage of patients with HF will die suddenly, however many will deteriorate more slowly and it is possible to recognise the very end-stage of disease and to plan care with some advance warning.

In the community, patients entering end-stage disease in the UK should be registered with the Gold Standards Framework (GSF) to help with co-ordination and planning of care. Information relating to the GSF can be found at http://www.goldstandardsframework.nhs.uk and www.modern.nhs.uk/chd. Patients can be included if they have 2 or more of the following criteria: NYHA Class III or IV symptoms, thought to be in the last year of life, repeated hospital admissions with symptoms of heart failure, or if the patient has difficult physical / psychological symptoms despite optimal tolerated therapy.

Patients, family members and carers should have contact numbers for nursing, medical and any other services required throughout the 24 hour period.

It is often more difficult to diagnose the terminal phase of heart failure than cancer. Consider the following points:

- Try to establish consensus within the team about a patient's condition.
- It is often difficult to accept that deterioration does not represent failure by the health care team.
- Heart failure patients with advanced disease and acute deterioration may have reversible causes for the decline and achieve surprising improvement with medication changes
- If the chances of recovery are uncertain, share this with the patient and/or family.

Other generic tools to help provide a high standard of consistent, well documented care of the dying, such as the Liverpool Care Pathway can be used for end-stage heart failure patients. The Preferred Priorities for Care patient-held document facilitates choice of where the patient would want to be at the end of life and can allow open discussion between patient, carers and health professionals. In addition, consideration should also be given to any documented Advance Decisions to Refuse Treatment.

MANAGEMENT OF THE DYING HEART FAILURE PATIENT

Recognition of the dying phase

This can be difficult to ascertain in heart failure patients because there can be an initial response to corrective treatment. However, the following features indicate that the patient is likely to be in the last phase of illness:

- **Brittle fluid balance control with no identifiable reversible precipitant**
- **Diuretic resistance: failure to respond within 2 -3 days of appropriate change in diuretic treatment**
- **Sustained hypotension**
- **Unable to tolerate ACEI / ARB or beta-blockers**
- **Worsening renal dysfunction**
- **Resistant hyponatraemia**
- **Hypoalbuminaemia**

NB Heart failure patients with advanced disease and acute deterioration may achieve surprising improvement with medication changes. If there is uncertainty regarding the stage of illness, share this uncertainty with the patient and / or family. The following guidance, or even use of e.g. Liverpool Care Pathway documentation does not preclude other appropriate management, or change of direction if it becomes apparent that the patient is no longer dying. Daily review by the medical and nursing team is important.

Management of the dying phase

Achieving symptom control for patients in the last few days of life is greatly enhanced by having appropriate drugs readily available for managing common symptoms. This often involves **anticipatory prescribing** of drugs and might involve having such drugs as Midazolam, Diamorphine, Hyoscine Butylbromide and Haloperidol (plus water for injection) readily available in case they are required at short notice.

The Liverpool Care Pathway (LCP) for the dying is appropriate for use in patients with heart failure and has symptom control algorithms within it.

The triggers used to suggest that the patient is entering the last days of life and prompt the use of the LCP are that the multi-professional team agrees that the patient is dying and has at least **two** of the following features:

- Patient bed-bound
- Semi-comatose
- Only able to take sips of fluid
- Unable to take oral medication any further

As the patient becomes weaker and has difficulty swallowing, non-essential medications should be discontinued. Medication essential for symptom benefit can be converted to subcutaneous administration including analgesics, anxiolytics and Furosemide.

Unless the patient has a poor oral intake, diuretics should be continued. In general all other cardiac medications can be withdrawn. Stop intravenous hydration, as it may worsen fluid retention and there is no clear evidence that it helps symptom control. Discuss the rationale for this with the patient and family.

Furosemide administered subcutaneously can be useful when venous access is difficult or unwanted, although the following considerations should be taken:

- It can be given as one-off doses or as a continuous infusion via a syringe driver
- Compatibilities with other drugs are not currently known, and it is therefore not recommended to mix with other medication such as diamorphine or anxiolytic drugs
- The recommended diluent is sodium chloride, however this is dose dependent and Furosemide may be used undiluted.
- Patients with nickel allergy should have a "soft-set" or paediatric cannula inserted if a continuous infusion is needed.

- Care must be taken to inject into non-oedematous skin as absorption may be otherwise affected. The upper anterior chest wall is usually oedema-free even when there is gross peripheral oedema.

Assessment of symptoms and adjustment of medications as appropriate should occur regularly, including mouth, bladder and bowel care.

Routine blood tests and measurement of BP, pulse and temperature should be stopped. Cardio Pulmonary Resuscitation (CPR) status should be established and clearly documented – especially a "Do Not Attempt Resuscitation (DNAR)" decision.

Psychological support of the patient and their carers is very important and sensitive communication is of paramount importance including clear discussion of the aim of care. The patient's spiritual care requirements should also be assessed and addressed.

THE COMMONEST SYMPTOMS TO ADDRESS IN TERMINAL END-STAGE HEART FAILURE CARE

NAUSEA AND VOMITING

Haloperidol 1.5mg as a stat dose.

For patients needing more than one injection per 24 hours, an infusion of e.g. 1.5mg – 3mg over 24 hours subcutaneous infusion via syringe driver can be given

Levomepromazine 6.25mg as a stat dose.

For patients needing more than one injection per 24 hours, an infusion of e.g. 6.25mg – 12.5mg over 24 hours subcutaneous infusion via syringe driver can be given

RETAINED SECRETIONS IN THE UPPER RESPIRATORY TRACT

This is often a concern to the family and sometimes staff, but not distressing to the patient, and occurs when the patient is too weak to expectorate their own secretions. Changing position or raising the head of the bed may help. If the patient is semi-conscious, nursing in the coma position may help drainage of retained secretions.

If secretions persist and the family are still distressed despite reassurance, consider using **Glycopyrronium** 0.2mg - 0.4mg subcutaneously as a stat dose or 0.8mg – 2.4mg over 24 hours by subcutaneous infusion via syringe driver, or alternatively **Hyoscine Butylbromide** 60mg - 120mg over 24 hours by subcutaneous infusion via syringe driver.

BREATHLESSNESS

Diamorphine 1mg - 2.5mg subcutaneously, or if able to swallow, **Morphine** 2.5mg - 5mg every 6 hours if the patient is not already on strong opioids. If already on strong opioids, seek advice from Palliative Care Services regarding appropriate starting dose of diamorphine.

If opioids are effective, consider starting a 24 hour subcutaneous infusion of diamorphine via a syringe driver. The dose is calculated from the total dose of diamorphine and morphine in the previous 24 hours. NB. To calculate the equipotent dose of diamorphine for any given dose of oral morphine, divide the oral morphine dose by 3.

Example: A patient had 4 x 3mg doses of oral Morphine and 2 x 2.5mg doses of subcutaneous Diamorphine in the past 24 hours. This is equivalent to 4mg Diamorphine (oral morphine contribution) + 5mg Diamorphine (Diamorphine contribution) = 9mg diamorphine / 24hours via subcutaneous syringe driver.

PAIN

Again, **Diamorphine** 1mg - 2.5mg subcutaneously, or if able to swallow, **Morphine** 2.5mg - 5mg every 6 hours in the patient not already on strong opioids and titrate according to response and pain. If already on strong opioids, seek advice from Palliative Care Services regarding appropriate starting dose of diamorphine.

If patient requires frequent doses, consider subcutaneous infusion via syringe driver with dose of diamorphine dependent on requirements in previous 24 hours.

AGITATION / TERMINAL RESTLESSNESS

Firstly, exclude any potentially treatable precipitating factors such as urinary retention, faecal impaction, pain, uncomfortable position in bed, and address these appropriately.

If no treatable factor, consider **Midazolam** 2.5mg - 5mg subcutaneously every 4 hours. If repeated doses are required, consider commencing syringe driver with dose dependent upon previous 24 hours requirement.

REFERRAL TO SPECIALIST PALLIATIVE CARE SERVICES

Almost all patients with end-stage heart failure will require a supportive / palliative care approach with the aim of maximising their quality of life. This process requires shared decision-making between the patients, their carers and healthcare professionals. In most cases, the professionals already caring for them will be able to continue providing care.

Guidelines as described in this book are designed to help them in this. However, if the professionals caring for a patient are not able to manage a specific problem satisfactorily, then referral to a specialist palliative care service should be considered, bearing in mind the eligibility criteria for the local services.

People requiring specialist palliative care referral usually have one of more of the following problems:

- Recurrent hospital admissions for decompensated heart failure despite optimal medical treatment.
- Cases where there are difficult communication issues (such as coping with uncertainty, prognosis, and preferred place of death).
- Cases where there are difficulties in determining future care planning.
- Complex physical or psychological symptoms despite optimally-tolerated therapy.
- Practical support needed to allow dying at home.
- Carers with high-risk of difficulty in coming to terms with bereavement

Of course, each geographical area may have developed their own criteria for referral to Palliative Care, and the above is only suggested as a guide. Individual area guidance must first be applied.

REFERENCES AND FURTHER READING

Booth S, Wade R, Johnson MJ et al. The use of oxygen in the palliation of breathlessness. A report of the expert working group of the scientific committee of the association of palliative medicine. Respiratory Medicine. 2004; 98: 66–77.

British Heart Foundation (2007) – Implantable cardioverter defibrillators in patients who are reaching the end of life. A discussion document for health professionals.

Bruera E. Pharmacological interventions in cachexia and anorexia. In Oxford Textbook of Palliative Medicine, 3rd Edition 2004, p552-5606.

Clinical knowledge Summaries available via http://cks.library.nhs.uk/home

'Decisions relating to cardiopulmonary resuscitation' (2007). A joint statement from the British Medical Association, the Resuscitation Council (UK), Royal College of Nursing.

De Conno. Mouth care. In Oxford Textbook of Palliative Medicine. 3rd Edition 2004, p673-687

Department of Health (2008) End of life Care Strategy. Promoting high quality care for all adults at the end of life.
www.dh.gov.uk/en/Healthcare/IntegratedCare/Endoflifecare/index.htm

Gillam JL, Gillam DG. The assessment and implementation of mouth care in palliative care: a review. The Journal Royal Society Promotion of Health 2006; 126(1): 33-37

Gold Standards framework available via www.goldstandardsframework.nhs.uk and www.modern.nhs.uk/chd

Hurwitz B (1998) Clinical guidelines and the law. Negligence, discretion and judgement. Radcliffe Medical Press Ltd

Jennings A. Opioids for the palliation of breathlessness in terminal illness. Cochrane Database of Systematic Reviews 2001, Issue 3 http://www.cochrane.org/reviews/en/ab002066.html

Johnson MJ, McDonagh T, Harkness A, MacKay S, Dargie H. Morphine for breathlessness in chronic heart failure. European Journal of Heart Failure. 2002; 4: 753–756.

Johnson MJ. Management of end-stage cardiac failure. Post Graduate Medical Journal. 2007; 83: 395-401

Larkin PJ, Sykes NP, Centeno C. et al. The management of constipation in palliative care: clinical practice recommendations. Palliative Medicine 2008:22; 796-807

Liverpool Integrated Care Pathway available via http://www.mcpcil.org.uk/frontpage

Lloyd-Williams M. Diagnosis and treatment of depression in palliative care. European J Pall Care 2002; 9(5): 186-8

Macmillan GP Facilitators (2007) 'The Palliative Care Handbook. Advice on clinical management.' Sixth edition

Miles CL, Goodman ML, Wilkinson S. Laxatives for the management of constipation in palliative care patients (Systematic Review) 2006; Issue 4. The Cochrane Collaboration

Morice AH. British Thoracic Society recommendations for the management of cough in adults. Thorax 2006; 61:1-24

NHS End of Life Care Programme & the National Council for Palliative Care (2008) Advance Decisions to Refuse Treatment, A guide for health and social care professionals. http://www.adrtnhs.co.uk/pdf/EoLC_DOC.pdf

Oxberry SG, Johnson MJ. Review of the evidence for the management of dyspnoea in patients with chronic heart failure. Current Opinion in Supportive and Palliative Care 2008, 2:84–88

Preferred priorities for Care – downloaded from http://www.endoflifecareforadults.nhs.uk/eolc/ppc.htm

Pyatt JR, Trenbath D, Chester M and Connelly DT. The simultaneous use of a biventricular implantable cardioverter defibrillator (ICD) and transcutaneous electrical nerve stimulation (TENS) unit: implications for device interaction Europace (2003) 5, 91– 93 doi: 10.1 053/eupc.2002.0277

Strasser F. Pathophysiology of the anorexia/cachexia syndrome. In Oxford Textbook of Palliative Medicine, 3rd Edition 2004, p520-5337

The Joint North and North East Lincolnshire Palliative Care Group (2004) 'Management of Cancer Pain in Adults (including guidelines for administration of drugs)'

Wood GJ. Management of intractable nausea and vomiting in patients at the end of life. JAMA 2007; 298(10);1196-1207

Working Party of the Merseyside and Cheshire Specialist Palliative Care and Cardiac Clinical Networks (2005) Symptom control guidelines for patients with end-stage heart failure and criteria for referral to specialist palliative care. Available via http://www.cmcn.nhs.uk

World Health Organisation. Cancer Pain Relief (2e). World health Organisation, Geneva, 1996 p.74

1855633R0004

Printed in Great Britain
by Amazon.co.uk, Ltd.,
Marston Gate.